Richard Anthony Proctor

Half-Hours with the Stars

A Plain and Easy Guide to the Knowledge of the Constellations

ctor

rs

the Knowledge of

Australia, Japan

.com

HALF-HOURS WITH THE STARS

A PLAIN AND EASY GUIDE TO THE KNOWLEDGE OF THE CONSTELLATIONS

SHOWING, IN TWELVE MAPS, THE POSITION FOR THE UNITED STATES OF THE PRINCIPAL STAR
GROUPS NIGHT AFTER NIGHT THROUGHOUT THE YEAR, WITH INTRODUCTION
AND A SEPARATE EXPLANATION OF EACH MAP.

TRUE FOR EVERY YEAR

MAPS AND TEXT SPECIALLY PREPARED FOR AMERICAN STUDENTS

BY

RICHARD A. PROCTOR, F.R.A.S.

AUTHOR OF "HALF HOURS WITH THE TELESCOPE," "EASY STAR LESSONS," "A LARGER STAR ATLAS," AND THE ARTICLES ON
ASTRONOMY IN THE "AMERICAN CYCLOPÆDIA" AND THE "CYCLOPÆDIA BRITTANICA," ETC., ETC.

———

" Here I may sit and rightly spell
Of every star that Heav'n doth show."—MILTON.

The Heavens declare the glory of God ; and the Firmament showeth His handiwork.—PSALMS xix : 1.

———

NEW YORK AND LONDON
G. P. PUTNAM'S SONS
The Knickerbocker Press
1887

Richard Anthony Proctor

Half-Hours with the Stars
A Plain and Easy Guide to the Knowledge of the Constellations

ISBN/EAN: 9783337042707

Printed in Europe, USA, Canada, Australia, Japan

Cover: Foto ©berggeist007 / pixelio.de

More available books at **www.hansebooks.com**

INTRODUCTION ON THE USE OF THE MAPS.

It is very easy to gain a knowledge of the stars, if the learner sets to work in the proper man-ner. But he commonly meets with a difficulty at the outset of his task. He provides himself with a set of the ordinary star-maps, and then finds himself at a loss how to make use of them. Such maps tell him nothing of the position of the constellations *on the sky.* If he happen to recog-nize a constellation, then indeed his maps, if properly constructed, will tell him the names of the stars forming the constellation, and also he may be able to recognize a few of the neighboring con-stellations. But when he has done this he may meet with a new difficulty, even as respects this very constellation. For if he look for it again some months later, he will neither find it in its for-mer place nor will it present the same aspect,—if indeed it happen to be above the horizon at all.

It is clear, then, that what the learner wants is a set of maps specially constructed to show him in what part of the sky the constellations are to be looked for. He ought on any night of the year to be able to turn at once to the proper map, and in that map he ought to see at once what to look for, toward what point of the compass each visible constellation lies, and how high it is above the horizon. And, if possible (as the present work shows is the case), *one* map ought to suffice to exhibit the aspect of the whole heavens, in order that the beginner may not be confused by turn-ing from map to map, and trying to find out how each fits in with the others.

It is to fulfil these requirements that the present maps have been constructed. Each exhibits the aspect of the whole sky at a given day and hour. The circumference of the map represents the natural horizon, the middle of the map representing the part of the sky which lies imme-diately overhead. If the learner hold one of these maps over his head, so as to look vertically upwards at it, the different parts of the horizon marked in round the circumference being turned towards the proper compass points, he will see the same view of the heavens as he would if he were to lie on his back and look upwards at the sky, only that the map is a planisphere and the sky a hemisphere.

But although this illustration serves to indicate the nature of the maps, the actual mode of using them is more convenient.

Let it first be noted that properly speaking the maps have neither top, bottom, nor sides. Each map may be held with any part of the circumference downward : then the centre of the map is to be looked upon as the top for that part of the circumference. The portion of the map lying beneath the centre represents the portion of the sky lying between the point overhead and a certain part of the horizon—the part in fact corresponding to the particular part of the circum-ference which is turned downwards. Thus if on any night we wish to learn what are the stars towards the north, we look for the map corresponding to that night. At the hour named the stars toward the north will be those shown between the centre of the map and the top; and, of course, we hold the map upside down so as to bring the centre above the northern part of the circumference.

But this matter will be more clearly understood by comparing the account of any of the accompanying maps with the map itself.

Again, it must be noted that, although the maps are necessarily arranged in a certain order, there is in reality no first or last in the series. The map numbered I. follows the map numbered XII. in exactly the same manner that the latter follows the map numbered XI. The maps form a circular series, in fact.

The only reason for numbering the maps as at present, is that the map numbered I. happens to exhibit the aspect of the sky at a convenient hour on the night of January 1st. It will be found that the dates follow on with intervals of seven or eight days right round the year, the end of the year falling in the left-hand column of the table under Map I., while the beginning of the year is in the right-hand column of the same map.*

It will be seen at once that a map can always be found corresponding to a convenient hour on any night of the year. (In midsummer, on a few of the dates mentioned under the maps, night has not begun at the hour named.) On any date named under a map, the aspect of the sky two hours later than that named is that represented in the following map. Thus at eight o'clock in the evening of January 22d, the aspect of the stars is as shown in Map I. At ten o'clock on the same night the aspect of the sky is that shown in Map II., as a date under that map shows. Applying this rule to the few occasions on which the hour named is not available for observation (five or six in all out of ninety-six dates), the observer can manage as well for those occasions as for any others.

Next, as to finding the north point, or any point of the compass which will enable the observer to determine the rest. If he is only familiar with the aspect of those seven bright stars of the Great Bear which have been called the Dipper, Charles' Wain, (really " The Churl's Wain,") the Butcher's Cleaver, and by other names, he can always determine the north point by means of the two stars called the Pointers, since these seven stars never set. In the explanation of each map I have shown where the Great Bear is to be looked for on each night, the observer being assumed to have such a general knowledge of the direction of the compass-points, as will suffice for the purpose of finding so marked a collection of stars. Thus the pole-star is found, and for the purpose of such observations as are here considered, this star may be looked upon as marking the exact direction of the north.

Perhaps nothing further is required ; but if the observer prefer it he can determine the north point conveniently *at noon* by setting up a vertical stick in the sunlight and noting the direction in which the shadow lies. Once the observation has been made, he can note what objects (these should be distant) lie towards the different points of the compass, and from that time he can use the accompanying maps without any reference to the Great Bear and the Pointers.

It is worth noticing that the stars called the Guardians of the Pole form no bad time-piece when used with the aid of such maps as the present. They revolve round the pole once in twenty-four hours (less about four minutes), in a direction contrary to that of a clock's hands. But stars near the equator, whose motions are much more rapid, afford a yet better measure of time, if the direction of the south point is well determined.

* It may be mentioned in passing, that the dates have not been thrown in so as to fall regularly round the year, but correspond with the variations due to the earth's variable motion round the sun.

Of course, the observer who really wishes to become an astronomer will not rest satisfied by learning only the principal stars shown in these maps. By means of the regular star-maps, such as those of my School Star Atlas, he will be able to explore the depths of all the constellations, having once learned their position and general appearance from the accompanying maps. It will be well for the student to remember that the planets Venus, Mars, Jupiter, and Saturn will at times appear among the constellations here shown. Venus and Jupiter can always be recognized by their superior light, Mars and Saturn by the steadiness with which they shine. The almanac will always show when these planets and Mercury (often very bright in the clear skies of America) are above the horizon, and where they are situate. They never appear except among the zodiacal constellations.

For particulars and pictures of the different constellations, and other details associated with the study of the star-groupings, the reader is referred to my " Easy Star Lessons," published like the present maps by Messrs. PUTNAM'S SONS. I have to thank the proprietors of the *Scientific American* for permission to publish these maps, which originally appeared (though in a slightly different form) in the pages of that excellent magazine. The Latin names of the constellations included in the maps of this series are as follows :

THE LITTLE BEAR, URSA MINOR α, the *Pole Star* ; β, γ, *the Guardians*).
THE DRAGON, DRACO (α, *Thuban*).
KING CEPHEUS, CEPHEUS.
THE LADY IN THE CHAIR, CASSIOPEIA.
THE CHAMPION, PERSEUS (β, *Algol*, remarkable variable).
THE CHARIOTEER, AURIGA (α, *Capella*).
THE GREATER BEAR, URSA MAJOR (α, β, the *Pointers*).
THE HUNTING DOGS, CANES VENATICI (α, *Cor Caroli*).
QUEEN BERENICE'S HAIR, COMA BERENICES.
THE HERDSMAN, BOÖTES (α, *Arcturus*).
THE NORTHERN CROWN, CORONA BOREALIS.
THE SERPENT, SERPENS.
THE KNEELER, HERCULES.
THE LYRE, LYRA (α, *Vega*).
THE SWAN, CYGNUS (α, *Arided* ; β, *Albireo*).
THE WINGED HORSE, PEGASUS.
THE CHAINED LADY, ANDROMEDA.
THE TRIANGLES, TRIANGULA.
THE RAM, ARIES.
THE BULL, TAURUS (α, *Aldebaran* ; η, *Alcyone*, the chief *Pleiad*).
THE TWINS, GEMINI (α, *Castor* ; β, *Pollux*).
THE CRAB, CANCER (the cluster between γ and δ is the *Beehive*).
THE LION, LEO (α, *Regulus*).
THE VIRGIN, VIRGO (α, *Spica*).
THE SCALES, LIBRA.
THE SERPENT-HOLDER, OPHIUCHUS.
THE EAGLE, AQUILA (α, *Altair*).
THE DOLPHIN, DELPHINUS.

THE WATER CARRIER, AQUARIUS.
THE FISHES, PISCES.
THE SEA MONSTER, CETUS (*o, Mira*, remarkable variable).
THE RIVER, ERIDANUS.
THE GIANT HUNTER, ORION (*a, Betelgeux ; β, Rigel*).
THE LESSER DOG, CANIS MINOR (*a, Procyon*).
THE SEA SERPENT, HYDRA (*a, Alphard*).
THE CUP, CRATER (*a, Alkes*).
THE CROW, CORVUS.
THE SCORPION, SCORPIO (*a, Antares*).
THE ARCHER, SAGITTARIUS.
THE SEA-GOAT, CAPRICORNUS.
THE SOUTHERN FISH, PISCIS AUSTRALIS (*a, Fomalhaut*).
THE HARE, LEPUS.
THE DOVE, COLUMBA.
THE GREATER DOG, CANIS MAJOR, (*a, Sirius*).
THE SHIP, ARGO.
THE CENTAUR, CENTAURUS.

The following table exhibits the names of all the stars of the first three magnitudes to which astronomers have given names ; at least, all those whose names are in common use :

a Andromedæ, *Alpheratz*
β ——, *Mirach, Misar*
γ ——, *Almach*
a Aquarii, *Sadalmelik*
β ——, *Sadalsund*
δ ——, *Skat*
a Aquilæ, *Altair*
β ——, *Alshain*
γ ——, *Tarazed*
a Arietis, *Hamal*
β ——, *Sheraton*
γ ——, *Mesartim*
a Aurigæ, *Capella*
β ——, *Menkalinan*
a Bootis, *Arcturus*
β ——, *Nekkar*
ε ——, *Izar, Mizar, Mirach*
η ——, *Muphrid*
a Canum Ven., *Cor Caroli*
a Canis Majoris, *Sirius*
β ——, *Mirzam*
ε ——, *Adara*
a Canis Minoris, *Procyon*
β ——, *Gomeisa*
a³ Capricorni, *Secunda Giedi*
δ ——, *Deneb Algiedi*
a Cassiopeiæ, *Schedar*
β ——, *Chaph*
a Cephei, *Alderamin*
β ——, *Alphirk*
γ ——, *Errai*
a Ceti, *Menkar*
β ——, *Diphda*

ζ Ceti, *Baten Kaitos*
o ——, *Mira*
a Columbæ, *Phact*
a Coronæ Bor., *Alphecca*
a Corvi, *Alchiba*
δ ——, *Algores*
a Crateris, *Alkes*
a Cygni, *Arided, Deneb Adige*
β ——, *Albireo*
a Draconis, *Thuban*
β ——, *Alwaid*
γ ——, *Etanin*
β Eridani, *Cursa*
γ ——, *Rowrat*
a Geminorum, *Castor*
β ——, *Pollux*
γ ——, *Alhena*
δ ——, *Wasat*
ε ——, *Mebsuta*
a Herculis, *Ras Algethi*
β ——, *Kornephoros*
a Hydræ, *Al Fard, Cor Hydræ*
a Leonis, *Regulus, Cor Leonis*
β ——, *Deneb Alect, Denebola, Deneb*
γ ——, *Algieba*
δ ——, *Zosma*
a Leporis, *Arneb*
a Libræ, *Zuben el Genubi*
β ——, *Zuben el Chamali*
γ ——, *Zuben Hakrabi*
a Lyræ, *Vega*
β ——, *Sheliak*
γ ——, *Sulaphat*

a Ophiuchi, *Ras Alhague*
β ——, *Cebalrai*
a Orionis, *Betelgeux*
β ——, *Rigel*
γ ——, *Bellatrix*
δ ——, *Mintaka*
ε ——, *Alnilam*
a Pegasi, *Markab*
β ——, *Scheat*
γ ——, *Algenib*
ε ——, *Enif,*
ζ ——, *Homan*
a Persei, *Mirfak*
β ——, *Atgol*
a Piscis Aust., *Fomalhaut*
a Sagittarii, *Kaus Australis*
a Scorpionis, *Antares, Cor Scorpionis*
a Serpentis, *Unukalhai*
a Tauri, *Aldebaran*
β ——, *Nath*
η ——, *Alcyone* (Pleiad)
a Ursæ Majoris, *Dubhe*
β ——, *Merak*
γ ——, *Phecda*
δ ——, *Alioth*
ζ ——, *Mizar*
η ——, *Alkaid, Benetnasch*
ι ——, *Talitha*
a Ursæ Minoris, *Polaris*
β ——, *Kochab*
a Virginis, *Spica Azimech, Spica*
β ——, *Zavijava*
ε ——, *Vindemiatrix*

At 11 o'clock : Dec. 7.		At 9 o'clock : Jan. 7.
At 10½ o'clock : Dec. 15.	At 9½ o'clock : Dec. 30.	At 8½ o'clock : Jan. 14.
At 10 o'clock : Dec. 23.		At 8 o'clock : Jan. 22.

Stars of the first magnitude are eight-pointed ; second magnitude, six-pointed ; third magnitude, five-pointed ; fourth magnitude (a few), four-pointed ; fifth magnitude (very few), three-pointed. For star-names refer to page 4.

NIGHT SKY.—DECEMBER AND JANUARY.

The Great Bear (*Ursa Major*) is now rising well above the horizon, in the northeast, the Pointers about midway between north and northeast. A line from the Pole Star to the Guardians of the Pole is now in the position of the minute hand of a clock about 28 minutes past an hour. The Dragon (*Draco*) lies due north, curving round under the Little Bear, its head close to the horizon. Low down in the northwest is a part of the Swan (*Cygnus*). Higher up we see King Cepheus, his wife *Cassiopeia*, and their daughter *Andromeda* (the Seated Lady and Chained Lady, respectively), with the Rescuer (*Perseus*) nearly overhead. The Winged Horse is setting, his head close by the western horizon, and near the jar of the Water Bearer (*Aquarius*).

In the southwest is the Whale; and close by, the constellation *Pisces*, or the Fishes; above them the Ram (*Aries*), between which and *Andromeda* the Triangles can be seen.

In the south the River (*Eridanus*) makes now its best show. Its leading brilliant, *Achernar*, is, however, never seen in the United States. In the southeast the Great Dog, with the splendid Sirius (" which brightliest shines when laved of ocean's wave "), shows resplendently. Above is Orion now standing upright, treading on the Hare (*Lepus*) and facing the Bull (*Taurus*), now at its highest. The Dove (*Columba*) below the Hare is a modern and not very interesting constellation.

The Little Dog (*Canis Minor*) is on the east of Orion. In the east the Sea Serpent (*Hydra*) is rising, and due east a little higher we find *Cancer*, the Crab, (note the pretty cluster called the Beehive (*Praesepe*); above are the twins (*Gemini*), and above them the Charioteer (*Auriga*), with the bright *Capella*, nearly overhead.

The Lion is rising in the northeast, his heart star *Regulus* (α) being low down a little north of east.

Lastly, due north, high up, the absurd Giraffe (*Camelopardus*) stands proudly on his ridiculous head.

NIGHT SKY.—JANUARY AND FEBRUARY.

The Great Bear (*Ursa Major*) with its Dipper and Pointers, occupies the northeasterly mid-heaven. A line from the Pole Star (α of the Little Bear, *Ursa Minor*) to the Guardians, β and γ, lies in the position of the minute hand of a clock 23 minutes after an hour. The Camelopard (*Camelopardus*) is above. The Dragon (*Draco*), whose head is below the horizon, curves round the Little Bear to between the Guardians and the Pointers. In the northwest, fairly high up, we find *Cassiopeia*, the Seated Lady, and on her right, lower down, the inconspicuous constellation *Cepheus*. *Andromeda*, the Chained Lady, is on *Cassiopeia's* left. The Great Nebula will be noticed in the map—it is faintly visible to the naked eye. Above *Andromeda* is *Perseus*, the Rescuing Knight, and above him the Charioteer (*Auriga*), nearly overhead. On the left of *Andromeda* is *Aries*, the Ram, the small constellation the Triangles lying between them.

Toward the southwest, the Whale (*Cetus*) is beginning to set. The River (*Eridanus*) occupies the lower part of the southwesterly sky, and extends also to the mid-heavens in that direction. The Dove (*Columba*) is nearly due south, and at its best—which is not saying much. Above is the Hare (*Lepus*), on which *Orion* treads. The Giant now presents his noblest aspect—prince of all the constellations as he is. He faces the Bull (*Taurus*), known by the *Pleiades* and the bright *Aldebaran*.

Close by the poor Hare, on the left, leaps *Canis Major*, the Greater Dog, with the bright Sirius, which "bickers into green and emerald." The stern of the Star Ship (*Argo*) is nearing the south.

Very high in the southeast we find the Twins (*Gemini*), with the twin stars, *Castor* and *Pollux* (α and β); and below them the Little Dog (*Canis Minor*). The Sea Serpent (*Hydra*) is rearing its tall neck above the eastern horizon (by south), as if aiming either for the Little Dog or for the Crab (*Cancer*), now high up in the east, with its pretty Beehive cluster showing well in clear weather. The Lion (*Leo*) is due east, the Sickle (marked by the stars α, η, γ, μ and ε) being easily recognized.

Queen Berenice's Hair (*Coma Berenices*, not *Berenicis*, as often ignorantly given) is in the northeast. It used to mark the tip of the real Lion's tail, just as the stars of the Crab marked his head. The Hunting Dogs occupy the space between Berenice's Hair and the Great Bear.

MAP II.   NIGHT SKY.—JANUARY AND FEBRUARY.

At 11 o'clock : Jan. 7.
At 10½ o'clock : Jan. 14.
At 10 o'clock : Jan. 22.

At 9½ o'clock : Jan. 29.

At 9 o'clock : Feb. 6.
At 8½ o'clock : Feb. 14.
At 8 o'clock : Feb. 21.

Stars of the first magnitude are eight-pointed ; second magnitude, six-pointed ; third magnitude, five-pointed ; fourth magnitude (a few), four-pointed ; fifth magnitude (very few), three-pointed.   For star-names refer to page 4.

MAP III. NIGHT SKY.—FEBRUARY AND MARCH.

At 11 o'clock: Feb. 6.		At 9 o'clock: Mar. 8.
At 10½ o'clock: Feb. 14.	At 9½ o'clock: Mar. 1.	At 8½ o'clock: Mar. 16.
At 10 o'clock: Feb. 21.		At 8 o'clock: Mar. 23.

Stars of the first magnitude are eight-pointed; second magnitude, six-pointed; third magnitude, five-pointed; fourth magnitude (a few), four-pointed; fifth magnitude (very few), three-pointed. For star-names refer to page 4.

NIGHT SKY.—FEBRUARY AND MARCH.

The Great Bear (*Ursa Major*), with its Dipper and Pointers, is now high up in the northeastern sky. The Pointers direct us to the Pole Star, (α of the Little Bear *Ursa Minor*). A line from the Pole Star to the Guardians of the Pole (β and γ) lies in the position of the minute hand of a clock 18 minutes after an hour. The Dragon (*Draco*) extends from between the Bears to the horizon—east of north—where its head with its two bright eyes can be seen.

Cepheus is low down, somewhat to the west of north; his Queen (*Cassiopeia*) the Seated Lady, beside him (α and β mark the top rail of her chair's back); while above her lies the poor constellation *Camelopardus*, the Giraffe.

Andromeda, the Chained Lady, is in the northwest, low down—in fact, partly set; the Triangles and the Ram (*Aries*) beside her, toward the west. Above them is *Perseus*, the Rescuing Knight; and above him, somewhat to the west, the Charioteer (*Auriga*). The Bull (*Taurus*), with the *Pleiades* and the bright *Aldebaran*, is in the mid-heaven, due west; *Gemini*, the Twins, higher, and toward the southwest. *Orion*, below them, is already slanting toward "his grave, low down in the west"; beneath him the Hare, and in the southwest a part of the River (*Eridanus*).

Due south is a part of the Star Ship (*Argo*), beside which, low down, is the foolish Dove (*Columba*), while above leaps the Great Dog (*Canis Major*), with the splendid *Sirius*, chief of all the stars in the sky, marking his mouth.

High up, a little west of north, is the Little Dog (*Canis Minor*); and higher, a little east of north, the Crab (*Cancer*), the "dark constellation," as it was called of old, with the pretty cluster *Præsepe*, or the Beehive.

The Sea Serpent (*Hydra*) is rearing his long neck high above the horizon, bearing on his back, absurdly enough, Noah's Cup (*Crater*) and Noah's Raven or Crow (*Corvus*).

Nearly due east, the Virgin (*Virgo*) has risen, Spica shining brightly just above the horizon. The Lion (*Leo*) occupies the mid-space above; the "Sickle in the Lion"—its handle marked by η and α, its curved blade by γ, μ, and ε—will at once be recognized. The Hair of Queen Berenice (*Coma Berenices*) is nearly due east, and fairly high. Between this small but remarkable group and the Great Bear, lies Hevelius's foolish constellation, the Hunting Dogs (*Canes Venatici*). Lastly, in the northeast, the Herdsman (*Boötes*), with the orange-yellow brilliant, Arcturus, is rising, though at present, paradoxical as it may seem, he lies on his back.

NIGHT SKY.—MARCH AND APRIL.

The Great Bear (*Ursa Major*) is now nearing the point overhead, the Pointers (*α* and *β*) aiming almost directly downward toward the Pole Star. The line from this star (*α* of the Little Bear, *Ursa Minor*) to the Guardians (*β* and *γ*) is now in the position of the minute hand of a clock about 13 minutes after an hour.

Cepheus lies north, low down, *Cassiopeia* on his left, the Camelopard above her, *Andromeda* just setting, almost due northwest, on the left. *Perseus* is due northwest, rather low, the Charioteer (*Auriga*) on his left, but higher. Setting between west and northwest we see the Bull (*Taurus*). with the *Pleiades* and the ruddy *Aldebaran*. *Orion* is almost prone in his descent toward his western grave. The Twins (*Gemini*) are due west, in the mid-heavens; the Little Dog (*Canis Minor*) beside them on their left, the Crab (*Cancer*) above, the Greater Dog (*Canis Major*) below, chasing the Hare (*Lepus*) below the horizon. Just behind the Dog the poop of the Great Ship (*Argo*) is also setting.

The Sea Serpent (*Hydra*) now shows his full length, rearing his head high in the south. Observe the darkness of the region around his heart, marked by the star *α*, *Alfard*, the Solitary One. The Cup (*Crater*) and Crow (*Corvus*) stand on his back.

The Sickle in the Lion (*Leo*) now stands with handle upright, due south. Below the tail stars of the Lion we see the Virgin (*Virgo*), with the bright *Spica Azimech*. The set of five third magnitude stars, above, was called by the Arabs, for reasons not explained, the " Retreat of the Howling She Dog."

Behind the Lion, due east and high up, we see *Coma Berenices*, the hair of Queen Berenice, between which and the tail of the Great Bear we see in the chart one star only of the Hunting Dogs (*Canes Venatici*).

The Herdsman (*Boötes*), still on his back, pursues in that striking and effective position the Great Bear. Below the shoulder stars of the Herdsman we see the Crown (*Corona Borealis*), near which, on the right, low down and due east, the head of the Serpent (*Serpens*) is rising. *Hercules* is also rising, but in the northeast.

Lastly, the stars of the Dragon (*Draco*) can be seen curving from between the Pointers and the Pole, round the Little Bear, then back toward *Hercules*, the head of the Dragon, with the bright eyes, *β* and *γ*, being rather low down, and somewhat north of northeast.

At 11 o'clock : Mar. 8.
At 10½ o'clock : Mar. 16.
At 10 o'clock : Mar. 23.

At 9½ o'clock : Mar. 30.

At 9 o'clock : Apr. 7.
At 8½ o'clock : Apr. 14.
At 8 o'clock : Apr. 22.

Stars of the first magnitude are eight-pointed ; second magnitude, six-pointed ; third magnitude, five-pointed , fourth magnitude (a few), four-pointed ; fifth magnitude (very few), three-pointed.   For star-names refer to page 4.

MAP V. NIGHT SKY.—APRIL AND MAY.

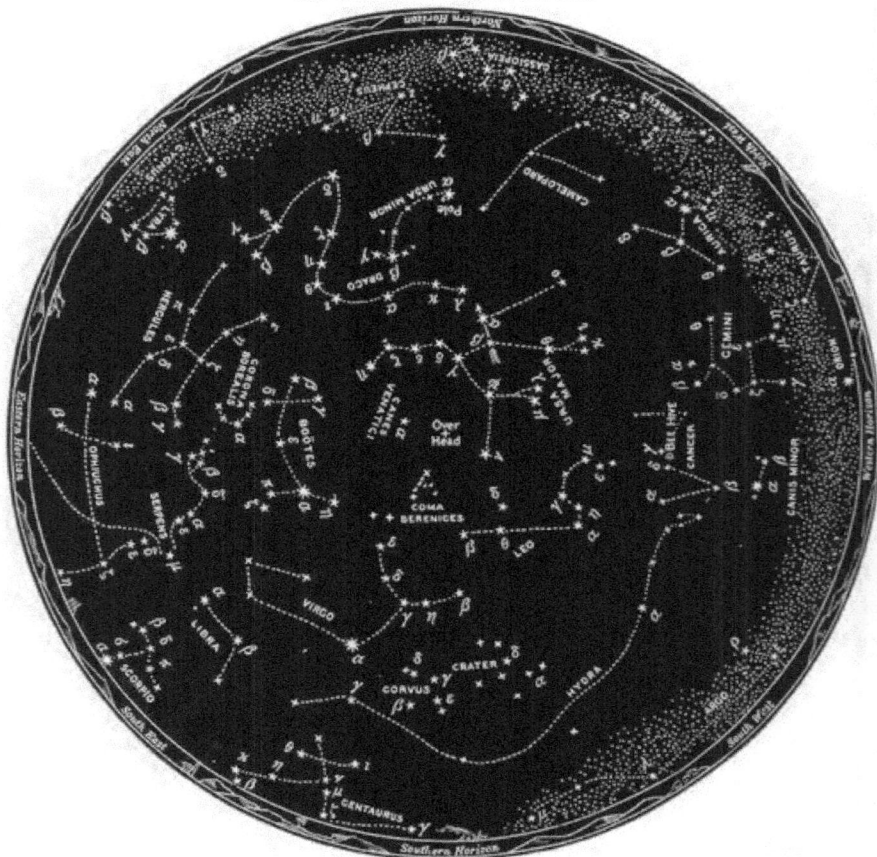

NIGHT SKY.—APRIL AND MAY.

The Great Bear (*Ursa Major*) is now at its highest and nearly overhead, the pointers aiming downward from high up, slightly west of due north. A line from the Pole Star, (*a* of the Little Bear, *Ursa Minor*) to the Guardians of the Pole, (*β* and *γ*) is now in the position of the minute hand of a clock 8 minutes after an hour.

Below the Little Bear we find *Cepheus* low down to the east of north, and *Cassiopeia* low down to the west of north. *Perseus*, the Rescuer, is setting in the northwest; the Camelopard is above, trying to get on his feet.

The Charioteer (*Auriga*), with the bright *Capella*, is nearing the northwestern horizon, followed by the Twins (*Gemini*), in the west. Further west and higher we find the Crab (*Cancer*), below which is the Little Dog (*Canis Minor*).

The southwestern sky is very barren of bright stars. *Alfard*, the heart of the Sea Serpent, *Hydra*, shines here alone in a great blank space. Above the Sea Serpent's head we see the Sickle in the Lion, *Leo* himself stretching his tail to due south, very high up. *Coma Berenices* is close by, and the Hunting Dogs (*Canes Venatici*) between *Coma* and the Great Bear.

In the south, lower down, we find the Crow (*Corvus*), and the Cup (*Crater*), on the Serpent's back; the Virgin (*Virgo*), extending in the mid-heavens from southeast to south, between the Lion's tail and the Crow. In the same direction, but low down, we find the head and body of the Centaur (*Centaurus*), supposed to have typified the patriarchal Noah.

In the southeast the Scorpion's heart has just risen, and between the head of *Scorpio* and the Virgin's robes we see the stars of the Scales (*Libra*).

Due east, low down, is the Serpent-Holder (*Ophiuchus*), on his back—it is the customary attitude of heavenly bodies when rising. The Serpent (*Serpens*) held by him is seen curving upward toward the Crown (*Corona Borealis*). The Serpent's head is due west, and above it we see the bright Arcturus, chief brilliant of the Herdsman (*Boötes*).

In the northeast is *Hercules*, his head close to the head of the Serpent-Holder. Beneath his feet is the Lyre (*Lyra*) with the brilliant *Vega*; and the Swan (*Cygnus*) has already half risen above the northeastern horizon.

Lastly, the Dragon (*Draco*) curves from between the Pointers and the Pole, round the Guardians toward *Cepheus*, and then retorts its head, with gleaming eyes (*β* and *γ*), toward the heel of *Hercules*.

NIGHT SKY.—MAY AND JUNE.

The Great Bear (*Ursa Major*) occupies all the upper sky from the west to north, except a small space occupied by the Hunting Dogs (*Canes Venatici*). The Pointers are in the northwest, almost horizontal. A line from the Pole Star (α of the Little Bear—*Ursa Minor*) to the Guardians of the Pole (β and γ) now occupies the position of the minute hand of a clock 3 minutes past an hour.

Due south, low down, lies *Cassiopeia*, while above, somewhat toward the east, we find the inconspicuous constellation *Cepheus*. The Camelopard is in the west of north, and getting upright.

Low down in the northwest lie the Charioteer (*Auriga*), and the head stars of the Twins (*Gemini*) further west. The Crab (*Cancer*) is nearly due west, the Sea Serpent (*Hydra*) holding his head almost exactly to the west point. Above is the Sickle in the Lion, its blade curved downward, and the tail of the Lion (*Leo*) lies above, toward the south of west.

On the Serpent's back we find the Cup (*Crater*) and the Crow (*Corvus*), in the southwest and to the south of southwest respectively. Above these constellations, and extending beyond the south toward the east, the Virgin (*Virgo*) occupies the mid-heavens.

Above the Virgin we see the Herdsman (*Boötes*), his head and shoulders nearly overhead. Low down in the south is the Centaur (*Centaurus*), bearing on his spear the Wolf (*Lupus*) as an offering for the Altar (*Ara*), which, however, is invisible in these latitudes. Above the Wolf we see the Scales (*Libra*), while the Scorpion (*Scorpio*), one of the few constellations which can at once be recognized by its shape, is rising balefully in the southeast.

The Serpent Bearer (*Ophiuchus*) bears the Serpent (*Serpens*) in the mid-heavens toward the southeast, the Crown (*Corona Borealis*) being high up in the east, close by the Serpent's head.

Low down in the east is the Eagle (*Aquila*), with the fine steel-blue star *Altair*, the Swan on the left about northeast, and above it the Lyre (*Lyra*), with the still more brilliant steel-blue star *Vega*. Hercules occupies the space between the Lyre on the one side and the Crown and the Serpent's head on the other. He is high up, due east.

Lastly, the Dragon winds from between the Pointers and the Pole round the Little Bear, toward Cepheus, and then eastward toward the feet of Hercules, close by which we see his head and gleaming eyes (β and γ).

At 11 o'clock: May 7.
At 10¼ o'clock: May 15.
At 10 o'clock: May 22.

At 9¼ o'clock : May 30.

At 9 o'clock : June 7.
At 8¼ o'clock: June 14.
At 8 o'clock: June 22.

Stars of the first magnitude are eight-pointed ; second magnitude, six-pointed ; third magnitude, five-pointed ; fourth magnitude (a few), four-pointed ; fifth magnitude (very few), three-pointed.   For star-names refer to page 4

At 11 o'clock : June 7.
At 10½ o'clock : June 14.   At 9½ o'clock : June 30.
At 10 o'clock : June 22.

At 9 o'clock : July 7.
At 8½ o'clock : July 14.
At 8 o'clock : July 22.

Stars of the first magnitude are eight-pointed ; second magnitude, six-pointed ; third magnitude, five-pointed ; fourth magnitude (a few), four-pointed ; fifth magnitude (very few), three-pointed.   For star-names refer to page 4.

NIGHT SKY.—JUNE AND JULY.

The Great Bear (*Ursa Major*) is in the mid-heavens toward the northwest, the Pointers not far from the horizontal position. They direct us to the Pole Star (α of the Little Bear, *Ursa Minor*). The line from this star to the Guardians of the Pole, β and γ, is in about the position of the minute hand of a clock 2 minutes before an hour. The Dragon (*Draco*) curls over the Little Bear, curving upward on the east, to where its head, high up in the northeast, is marked by the gleaming eyes, β and γ. Under the Little Bear, the Camelopard has at last come upright.

Low down in the west the Lion (*Leo*) is setting. The point of the " Sickle in the Lion " is turned toward the horizon ; the handle (marked by α and η) is nearly horizontal. Above the Lion's tail is Berenice's Hair (*Coma Berenices*); and between that and the Great Bear's tail our chart shows a solitary star of the Hunting Dogs (*Canes Venatici*). The Crow (*Corvus*) is low down in the southwest, the Cup (*Crater*) beside it, partly set, on the right. Above is *Virgo*, the Virgin. Still higher in the southwest—in fact, with head close to the point overhead—is the Herdsman (*Boötes*), the Crown (*Corona Borealis*) near his southern shoulder marking what was once the Herdsman's uplifted arm.

Low down between the south and southwest we find the head and shoulders of the Centaur (*Centaurus*), who holds the Wolf (*Lupus*) due south. Above the Wolf are the Scales (*Libra*), and above these the Serpent (*Serpens*), his head in the south, stretching toward the Crown. In the midsky, toward the southeast, we find the Serpent Bearer (*Ophiuchus*—one star of the Serpent lies east of him). Below the Serpent Bearer we find the Scorpion (*Scorpio*), now fully risen, and showing truly scorpionic form. Beside the Scorpion is the Archer (*Sagittarius*), low down in the southeast. To his left we see, low down, two stars marking the head of the Sea Goat (*Capricornus*), and one belonging to the Water Bearer (*Aquarius*). Above the Sea Goat flies the Eagle (*Aquila*), with the bright star *Altair* ; and above, near the point overhead, is the kneeling *Hercules*. Due east, we see part of the Winged Horse (*Pegasus*); above that, the little Dolphin (*Delphinus*), and higher, the Swan (*Cygnus*) and the Lyre (*Lyra*), with the beautiful bluish-white star *Vega*.

Lastly, low down, between north and northeast, we find the Seated Lady (*Cassiopeia*) ; and above, somewhat eastwardly, the inconspicuous constellation *Cepheus*, *Cassiopeia's* royal husband.

NIGHT SKY.—JULY AND AUGUST.

The Great Bear (*Ursa Major*) is now in the northwest, his paws near the horizon. The Pointers (*α* and *β*) direct us to the Pole Star, (*α* of the Little Bear, *Ursa Minor*). A line from the Pole Star to the Guardians of the Pole is in the position of the minute hand of a clock about 7 minutes before an hour. Below the Little Bear we see the Camelopard, a little to the east of due north. The Dragon (*Draco*) curves round from between the Pointers and the Pole, above the Little Bear toward the east, then upward to near the point overhead, its head, with the bright stars *β* and *γ*, being highest. Low down in the west we see Berenice's Hair (*Coma Berenices*), and one star of the Hunting Dogs (*Canes Venatici*) is seen in the chart between *Coma* and the Great Bear. The Herdsman (*Boötes*) occupies the mid-heaven in the west, the Crown (*Corona Borealis*) higher up, and due west, Hercules, between the Crown and the point overhead.

Low down, extending from the west to near the southwest, we find the Virgin (*Virgo*), the bright *Spica* near its setting place. In the southwest are the Scales (*Libra*), and farther to the left, extending from the Scales to low down near the south, we find the Scorpion (*Scorpio*), one of the finest of the constellations, *Antares*, the rival of Mars (as the name means), marking its heart. Above the Scorpion and the Scales are the Serpent Bearer (*Serpentarius* or *Ophiuchus*) and the Serpent (*Serpens*), extending right across him to near the Crown, after which the Serpent seems reaching.

A little east due south, low down, we find the Archer (*Sagittarius*); in the southeast, low down, the Sea Goat (*Capricornus*); and farther east, and lower down, the Water Bearer (*Aquarius*). Above the Sea Goat is the Eagle (*Aquila*), with the bright bluish-white star *Altair*; on its left the pretty little Dolphin (*Delphinus*), and above the Dolphin, nearly overhead, the Lyre (*Lyra*), with the bluish-white star *Vega* (even brighter than *Altair*) nearly overhead.

Below the Lyre we see the Swan (*Cygnus*), due east; and below the Swan the Winged Horse (*Pegasus*), upside down, as usual.

In the northeast, *Andromeda*, the Chained Lady, is rising, her head marked by the star *α* (which was also called *δ* of *Pegasus*. (The "Square of *Pegasus*" is formed by *α* of *Andromeda* and *α*, *β*, and *γ* of *Pegasus*.)

Between the north and northeast is *Cassiopeia*, the Seated Lady, and above her, her husband, King *Cepheus*. And lastly, *Perseus* is just rising, between the north and northeast.

MAP VIII. NIGHT SKY.—JULY AND AUGUST.

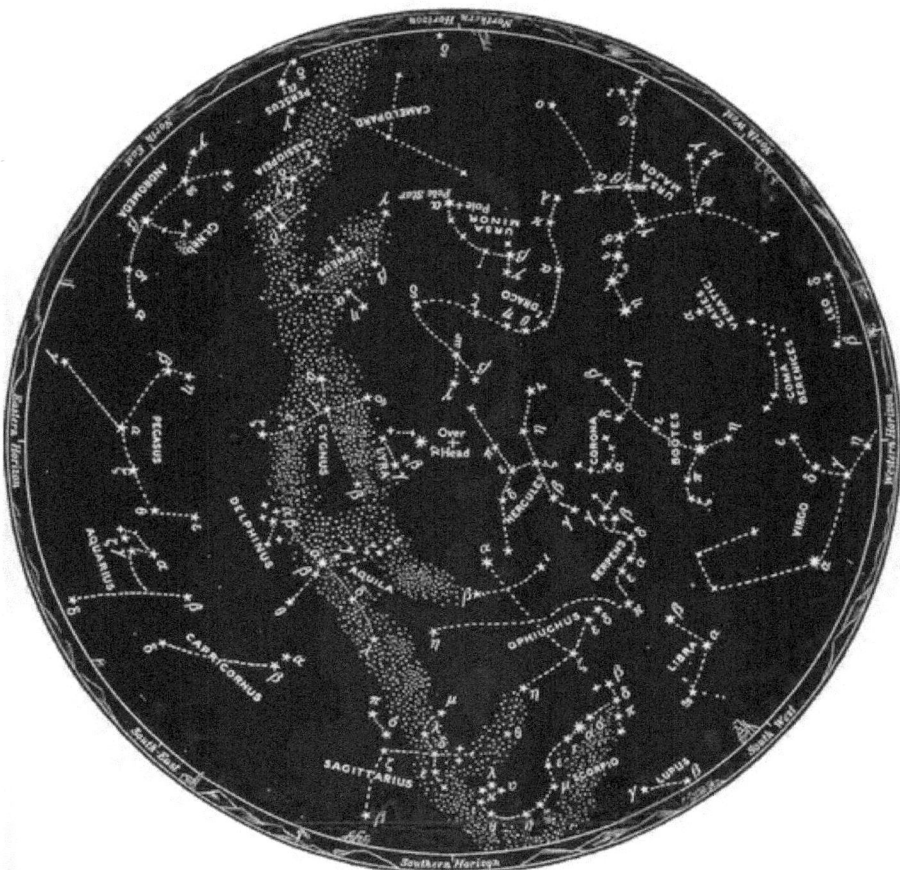

At 11 o'clock : July 7.
At 10½ o'clock : July 14.   At 9½ o'clock : July 30.
At 10 o'clock : July 22.

At 9 o'clock : Aug. 7.
At 8½ o'clock : Aug. 14.
At 8 o'clock : Aug. 22.

Stars of the first magnitude are eight-pointed ; second magnitude, six-pointed ; third magnitude, five-pointed ; fourth magnitude (a few), four-pointed ; fifth magnitude (very few), three-pointed. For star-names refer to page 4.

MAP IX. NIGHT SKY.—AUGUST AND SEPTEMBER.

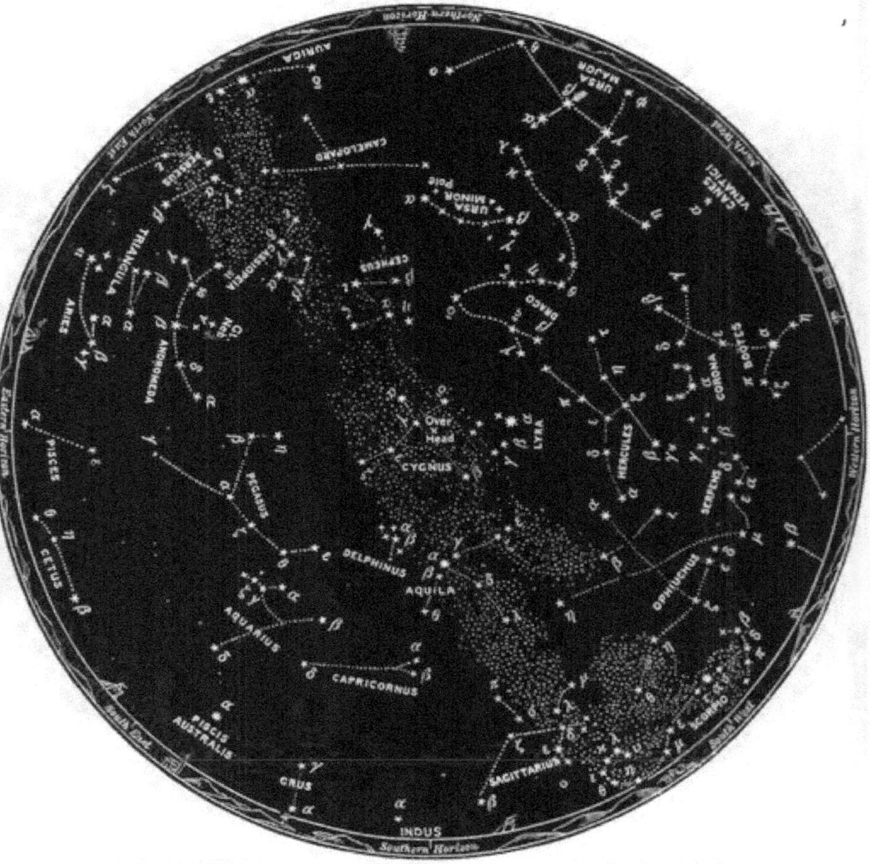

At 11 o'clock : Aug. 7.
At 10½ o'clock : Aug. 14.
At 10 o'clock : Aug. 22.

At 9½ o'clock : Aug. 29.

At 9 o'clock : Sept. 6.
At 8½ o'clock : Sept. 14.
At 8 o'clock : Sept. 21.

Stars of the first magnitude are eight-pointed ; second magnitude, six-pointed ; third magnitude, five-pointed ; fourth magnitude (a few), four-pointed ; fifth magnitude (very few), three-pointed.   For star-names refer to page 4.

NIGHT SKY.—AUGUST AND SEPTEMBER.

The Great Bear (*Ursa Major*) is low down, between northwest and north, the Pointers (α and β) directed slantingly upward toward the Pole. A line from the Pole Star (α of the Little Bear, *Ursa Minor*) to the Guardians of the Pole (β and γ), is in the position of the minute hand of a clock 12 minutes before an hour. Between the Great Bear and the Little Bear run the stars of the Dragon (*Draco*), round the Little Bear toward the north, thence toward the northwest, where we see the head of the Dragon high up, its two bright eyes, β and γ, directed toward *Hercules*, which occupies the western mid-heaven. Above Hercules is *Lyra*, the Lyre, with the bright steel-blue star Vega high up toward the point overhead. Right overhead is the Swan (*Cygnus*).

Low down in the northwest we see in the chart one star of the Hunting Dogs (*Canes Venatici*). Nearer the west stands the Herdsman, rather slanting forward, however, with the Crown (*Corona Borealis*) on his left, almost due west. The long winding Serpent (*Serpens*) runs from near the Crown (where we see its head due west) to farther south than southwest, high up on the western side of the Serpent Holder (*Serpentarius* or *Ophiuchus*), now standing upright in the southwest. Low down creeps the Scorpion (*Scorpio*), its heart Antares, rival of Mars, in the southwest, the end of its tail between south and southwest. Above and south of the Scorpion's tail we see the Archer (*Sagittarius*).

Due south, and high up, is the Eagle (*Aquila*), its tail at ζ and ε, its head at θ, the bright steel-blue Altair marking its body. On the left, or east, of the Eagle lies the neat little Dolphin (*Delphinus*). Midway between the Dolphin and the horizon is the tip of the tail of the Sea Goat (*Capricornus*), whose head lies nearly due south.

On the southern horizon is the head of the Indian (*Indus*); on its left a part of the Crane (*Grus*), and low down in the southeast lies Fomalhaut, the chief brilliant of the Southern Fish (*Piscis Australis*). Above lies the Water Bearer (*Aquarius*), in the southwestern mid-heaven.

Due east, fairly high, is "the Square of Pegasus," the head of the Winged Horse, Pegasus lying close by the Water Pitcher of Aquarius (marked by the stars ζ, γ, and α).

The Fishes (*Pisces*) are low down in the east. A few stars of the Whale (*Cetus*) are seen on their right, very low down. On the left of Pisces we see the Ram (*Aries*), low down; above it the Triangle; and above that the Chained Lady (*Andromeda*).

Low down in the northeast is the Rescuing Knight (*Perseus*); above whom is *Cassiopeia*; and on her left, higher up, the inconspicuous constellation *Cepheus*.

Lastly, immediately below *Cepheus*, we find the Camelopard, below which, very low down, between north and northeast, is the Charioteer (*Auriga*), the brilliant Capella being just above the horizon.

NIGHT SKY.—SEPTEMBER AND OCTOBER.

Low down, between north and nothwest, we find the seven stars of the Dipper, the Pointers on the right nearly due north. They direct us to the Pole Star. The Guardians of the Pole (β and γ of the Little Bear, *Ursa Minor*) lie in a direction from the Pole Star corresponding to that of the minute hand of a clock about 17 minutes before an hour. Between the Pointers and the Pole Star we find the tip of the Dragon's tail: then passing round the Little Bear with the Dragon's long train of third magnitude stars, we come, after a bend, to the Dragon's head, with the two bright eyes, α and β—(part of the Dragon's nose has been borrowed by Hercules). These two stars are almost exactly midway between the horizon and the point overhead, and nearly northwest. King Cepheus —not a very conspicuous constellation—lies between the point overhead and the Little Bear.

Low down in the northwest we find the head of the Herdsman (*Boötes*). The Crown (*Corona Borealis*), which no one can mistake, lies on his left; and close by is the setting head of the Serpent. Above these three groups we see Hercules—the Kneeler—his head at α, his upraised club by γ. Above the head of Hercules we find the Lyre, with the bright star Vega; and above that the Swan.

Passing southward, we see the Serpent-Holder (*Serpentarius* or *Ophiuchus*), beyond whom lies the Serpent's tail; a most inconvenient arrangement, as the Serpent is divided into two parts. Almost exactly southeast, and low down, are the stars of the Archer (*Sagittarius*); while above, in the mid-sky, we see the Eagle (*Aquila*), with the bright Altair. Note the neat little constellation the Dolphin (*Delphinus*), close by.

Due south is the Crane (*Grus*); above it the Southern Fish, with the bright star Fomalhaut. Above that the Sea Goat (*Capricornus*), and on the left of this the Water Bearer (*Aquarius*); one can recognize his water pitcher, marked by the stars β, γ, and α.

Toward the west, high up, is the Winged Horse (*Pegasus*); he is upside down just now. Below lies the Whale (*Cetus*), or rather the Sea Monster. I have my own notion about Cetus, regarding him as an icthyosaurus: but that is neither here nor there. The star o of this constellation is called Mira; it is a wonderful variable star. The Fishes (*Pisces*) may be seen between the Whale and Pegasus. Few constellations have suffered more than Pisces by the breaking up of star groups. The Fishes themselves are now lost in Andromeda and Pegasus.

Note how on the left of Pisces the Ram (*Aries*) "bears aloft" Andromeda, the Chained Lady (whose head lies at α), as Milton set Aries doing long since. The Triangles serve only as a saddle. Between Andromeda and her father, Cepheus, we find her mother, Cassiopeia, or rather Cassiopeia's Chair. (Of course β, α, and γ mark the chair's back.) Perseus, the Rescuer, lies below; β is the famous variable Algol. Below him lies the Bull (*Taurus*), with the Pleiades and the bright Aldebaran. Low down to the left of the Bull, we find the Charioteer (*Auriga*), with the bright Capella. And lastly, any one who likes may admire the Camelopard (*Camelopardalis*), between the Great Bear, Cepheus, and the Charioteer.

MAP X. NIGHT SKY.—SEPTEMBER AND OCTOBER.

At 11 o'clock : Sept. 6.
At 10½ o'clock : Sept. 14.
At 10 o'clock : Sept. 21.

At 9½ o'clock : Sept. 29.

At 9 o'clock : Oct. 7.
At 8½ o'clock : Oct. 15.
At 8 o'clock : Oct. 22.

Stars of the first magnitude are eight-pointed ; second magnitude, six-pointed ; third magnitude, five-pointed ; fourth magnitude
(a few), four-pointed ; fifth magnitude (very few), three-pointed. For star-names refer to page 4.

At 11 o'clock : Oct. 7.
At 10½ o'clock : Oct. 15.
At 10 o'clock : Oct. 22.

At 9½ o'clock : Oct. 30.

At 9 o'clock : Nov. 7.
At 8½ o'clock : Nov. 14.
At 8 o'clock : Nov. 22.

Stars of the first magnitude are eight-pointed ; second magnitude, six-pointed ; third magnitude, five-pointed ; fourth magnitude (a few), four-pointed ; fifth magnitude (very few), three-pointed.   For star-names refer to page 4.

NIGHT SKY.—OCTOBER AND NOVEMBER.

The Dipper lies low, the Pointers a little east of north. They direct to the Pole Star. The Guardians of the Pole (β and γ of the Little Bear, *Ursa Minor*) lie in a direction from the pole star corresponding to that of the minute hand of a clock about 22 minutes before an hour. Between the Pointers and Pole Star lies the tip of the Dragon's Tail. Sweeping around the Little Bear (*Ursa Minor*) we find the stars of the Dragon (*Draco*) curving back by the star δ to the Dragon's Head, with the two bright eyes, γ and β. Above is the inconspicuous constellation Cepheus; and somewhat higher, the stars of Cassiopeia, α and β, marking the top rail of the Seated Lady's Chair.

Low down in the northwest Hercules is setting. Above is the Lyre, with the bright steel-blue Vega; and above that the stars of the Swan (*Cygnus*), which has sometimes been called the Northern Cross.

Nearly due west we find the Eagle (*Aquila*), ζ and ϵ marking its tail, θ the head. Above the Eagle is the pretty little constellation *Delphinus*, the Dolphin.

In the southwest, rather low, is the Sea Goat (*Capricornus*); above and to the south of him the Water Bearer (*Aquarius*), with his pitcher, marked by the stars, α, γ, and ζ. The head of the Winged Horse, *Pegasus*, now upside down (in fact, he is seldom otherwise), is just above this group. The "Square of Pegasus" will be noticed high up, due south. The star α of Andromeda, one of the corners of this square, used to be also called δ of Pegasus.

Much attention need not be directed to the Phœnix, low in the southern horizon. The River *Eridanus* is coming well into view; and the great Sea Monster (*Cetus*) now shows finely, his head at α and γ, his paddles at ζ and τ. The Fishes (*Pisces*) are above: the Ram (*Aries*) above them and eastward, lying toward the southeast; then the Triangle (*Triangula*, or the Triangles, according to modern maps), and the Chained Lady (*Andromeda*) too nearly overhead to be very pleasantly observed. The great nebula in which the new star recently appeared is near the point overhead.

The grand giant Orion is rising in the east; above him the Bull (*Taurus*) with the Pleiades. Low down in the northeast the Twins (*Gemini*) are rising; above is the Charioteer (*Auriga*), and above him the Rescuing Knight (*Perseus*), "of fair-haired Danae born." The Camelopard is hardly worth noticing, except as marking a barren region of the heavens.

NIGHT SKY.—NOVEMBER AND DECEMBER.

The Great Bear (*Ursa Major*) is beginning to rise above the northeast (by north) horizon. The end of the Dipper's handle is hidden. A line from the Pole Star (toward which the Pointers direct the observer) to the Guardians of the pole (β and γ of the Little Bear, *Ursa Minor*), is now in the position of the minute hand of a clock 27 minutes before an hour. The stars of the Dragon wind round below the Little Bear toward the west, the head of the Dragon with the gleaming eyes (" oblique retorted that askant cast gleaming fire ") being low down, a little north of northwest. Above is King Cepheus, and above him his queen, the Seated Lady (*Cassiopeia*); their daughter, the Chained Lady (*Andromeda*) being nearly overhead.

Low down in the northwest we see the Lyre (*Lyra*), with the bright Vega; and close by toward the west the Swan (*Cygnus*), or Northern Cross. The Eagle is setting in the west, and the Little Dolphin nears the western horizon.

Toward the southwest (by west) we see the Water Bearer (*Aquarius*), with his pitcher (β, γ, α), close by which is the head of the Winged Horse (*Pegasus*). In the south, low down, is the absurd Phœnix; above, the Sea Monster, or Whale (*Cetus*); above him, the Fishes (*Pisces*); above them, the Ram (*Aries*); while nearly overhead lies the Triangle, in reality the Triangles (*Triangula*).

The River (*Eridanus*) occupies the southeasterly sky. The Dove and Great Dog (*Columba* and *Canis Major*) are rising in the southeast. The glorious *Orion* has now come well into view, though not yet so upright as we could wish a knightly hunter to be. He treads on the Hare (*Lepus*), and faces the Bull (*Taurus*) above.

Due east we find the Crab (*Cancer*), and Little Dog (*Canis Minor*) low down; the Twins (*Gemini*) higher; above them the Charioteer (*Auriga*), with the bright *Capella*, and *Perseus* the Rescuer nearing the point overhead. In the midspace between *Perseus*, *Auriga*, and the two Bears, we find the ridiculous constellation *Camelopardus*, or the Giraffe.

MAP XII. NIGHT SKY.—NOVEMBER AND DECEMBER.

At 11 o'clock : Nov. 7.
At 10½ o'clock : Nov. 14.   At 9½ o'clock : Nov. 30.
At 10 o'clock : Nov. 22.

At 9 o'clock : Dec. 7.
At 8½ o'clock : Dec. 15.
At 8 o'clock : Dec. 23.

Stars of the first magnitude are eight-pointed ; second magnitude, six-pointed ; third magnitude, five-pointed ; fourth magnitude
(a few), four-pointed ; fifth magnitude (very few), three-pointed.   For star-names refer to page 4.